Copyright © 2016 by Paul Mark

All rights reserved. No part of this publication may be reproduced, distributed, or transmitted in any form or by any means, including photocopying, recording, or other electronic or mechanical methods, without the prior written permission of the publisher, except in the case of brief quotations embodied in critical reviews and certain other noncommercial uses permitted by copyright law. For permission requests and any kind suggestion, write to the

authorpaul.mark@yahoo.com

FIRST EDITION: 2016

VOLUME: I

BY: PAUL MARK

Preface

This book features various basic concept related to basic engineering thermodynamics in a simple and creative way with illustrative figures. The author mainly focuses on precise and concise model of studying engineering thermodynamics concept with ease.

I am full of gratitude to my parents for the patience shown and encouragement given to complete this venture.

Every effort has been made to avoid errors and mistakes; however their presence cannot be ruled out. Any suggestions to improve the standard of this book will be highly appreciated.

Contents

- Chapter – 1 : 4
- BASIC ENGINEERING THERMODYNAMICS 4
- Chapter – 2 : 20
- ENERGY AND ITS FORMS 20
- Chapter – 3 : 27
- FIRST LAW OF THERMODYNAMICS 27
- Chapter – 4 : 29
- PURE SUBSTANCES 29
- Chapter – 5 : 41
- CONCEPT OF BOUNDARY WORK 41
- Chapter – 6 : 48
- CONCEPT OF FLOW WORK 48
- Chapter – 7 : 51
- SECOND LAW OF THERMODYNAMICS 51
- Chapter – 8 : 60
- CARNOT CYCLE 60
- Chapter – 9 : 64
- ENTROPY 64

Chapter – 1

BASIC ENGINEERING THERMODYNAMICS

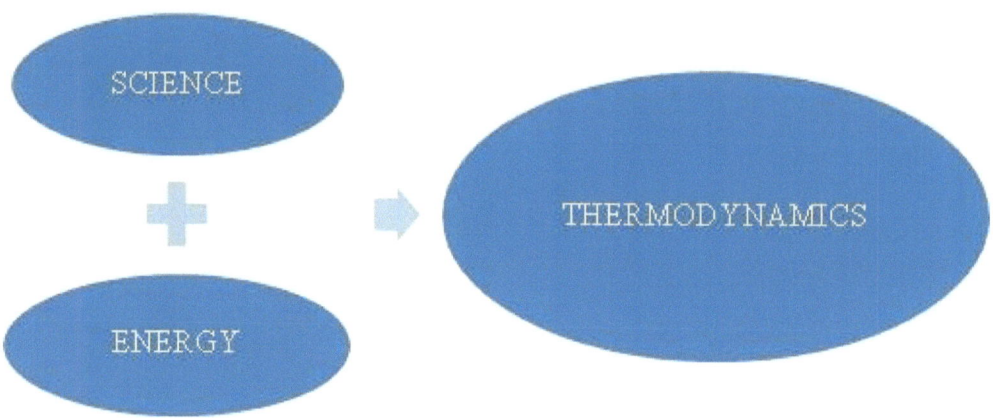

Thermodynamics is that branch of science in which we study about the interactions of energy mainly heat and work. In general, Thermodynamic term is derived from Greek words – therme (heat) and dynamis (power) which means convert heat into power. In other words, It is the study of the effects of work, heat, and energy on a system. But today, it mainly focuses on transformation of energies such as power generation, refrigeration, internal combustion engine and the relationship between the various properties of matter.

Demonstration: - If we like to cool our room on a hot humid day or rise the temperature of water in kettle then, What is the smallest amount of electricity or fuel required to achieve it? or if we would like to burn some coal or gas in a power plant to generate electricity, then

What is the largest amount of energy needed to get this? Only, Thermodynamics allows us to answer these types of questions.

IMPORTANT TERMINOLOGIES:-

Dimension and units: - Any physical quantity can be characterized by dimension and the magnitudes assigned to the dimension are called units. There have been basically two type of quantity – fundamental and derived quantity. Some example of fundamental quantity are Time, Mass, Length etc. and quantity like energy, acceleration, velocity are some example of derived quantity. In this book, we mainly deal with only metric S.I unit system which is accepted worldwide except some parts of USA. Although, there have some other system like English system (common units of this are pound, foot, gram etc.) but S.I. is a simple and logical system based on a decimal relationship between various units & it is being used by many international and national companies.

Fundamental Quantities	SI Unit	Symbol
Length	meter	m
Mass	kilogram	kg
Time	second	s
Electric current	Ampere	A
Temperature	Kelvin	K
Amount of subtance	mol	mol
Luminous Intensity	candela	cd

Some precaution: - The abbreviation of a unit was to capitalized if the unit was derived from a proper name (for example – S.I. unit of force which is named after sir I. Newton is newton not Newton and it is abbreviated as N). Also, the full name of a unit may be pluralized, but its abbreviation cannot.

Dimensional homogeneity: - In engineering science, all equation must be dimensionally homogeneous. That is, every term in an equation must have same unit. Dimension can be used to check formulas as well as they can be used to derive formulas. Also, a formula that is not dimensionally homogeneous is definitely wrong, but a dimensionally homogeneous formula is not necessary right.

SYSTEM AND CONTROL VOLUMES: -

System: - A system is defined as a quantity of matter or a region in space chosen for study or the part of the universe, with well-defined boundaries, one has chosen to study.

Surrounding: - The mass or region outside the system is called surrounding.

Boundary or wall: - The real or imaginary surface that separates the system from its surrounding is called boundary.

The boundary of a system can be fixed or movable. Note that the boundary is the contact surface shared by both the system and surroundings. Mathematically, the boundary has zero thickness and thus it can neither contain any mass nor occupy any volume in space.

TYPE OF THERMODYNAMIC SYSTEM: -

This is a system that could be described in terms of thermodynamic coordinates or properties. Thermodynamic Systems can be categorized into the followings depending on the type of boundary:

Open system: - This is a system in which boundary allows transfer of mass and energy into or out of the system. In other words, the boundary allows exchange of mass and energy between the system and the surrounding.

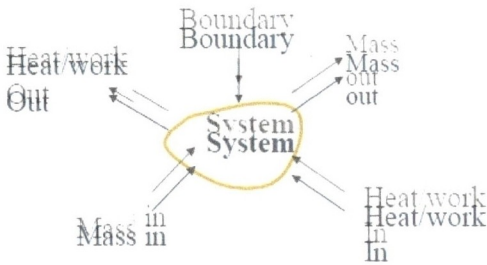

In other words, An open system or closed volume may be defined as selected region in space. Both mass and energy can cross the boundary of a control volume. The boundary of a control volume is called control surfaces and they can be real or imaginary. Most of the engineering devices are open system.

Closed system: - This is a system in which boundary allows exchange of energy alone (in form of heat) between the system and its surrounding (i.e. the boundary allows exchange of energy alone). This type of boundary that allows exchange of heat is called diathermal boundary.

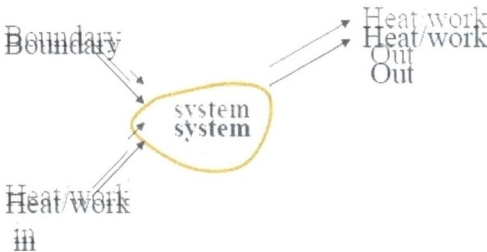

No mass entry or exit

In other words, A closed system consists of a fixed amount of mass and no mass can cross its boundary. But energy can cross the boundary and the volume of closed system does not have to be fixed. That is, we do permit heat and work to enter or leave but not mass.

Isolated system: - This is a system in which boundary allows neither mass nor energy to cross it. In other words, the boundary does not allow exchange of mass or energy.

In other words, It is of fixed mass and energy, and hence there is no mass and energy transfer across the system boundary.

CHOICE OF THE SYSTEM AND BOUNDARIES: -

We must choose the system for each and every case we work on, so as to obtain best possible output. We must be clear in defining what constitutes our system.

The boundaries may be real physical surfaces or they may be imaginary for the convenience of analysis.

Demonstration: - If the air in the room is the system, then the floor, ceiling and walls constitutes real boundaries and the plane at the open doorway constitutes an imaginary boundary.
The boundaries may be at rest or in motion.

Demonstration: - If we choose a system that has a certain defined quantity of mass (such as gas contained in a piston cylinder device) the boundaries must move in such way that they always enclose that particular quantity of mass if it changes shape or moves from one place to another.

MACROSCOPIC AND MICROSCOPIC APPROACHES: -

In macroscopic approach, certain quantity of matter is considered, without a concern on the events occurring at the molecular level. For example, temperature, pressure.

In microscopic approach, the effect of molecular motion is considered. For example, at microscopic level the pressure of a gas is not constant and the temperature of a gas is a function of the velocity of molecules.

Most microscopic properties cannot be measured with common instruments.

THERMODYNAMIC PROCESSES: -

A system undergoes a thermodynamic process when there is some sort of energetic change, temperature, or any sort of heat transfer.

There are several specific types of thermodynamic processes that happen frequently enough that they are commonly treated in the study of thermodynamics. Each has a unique trait that identifies it, and which is useful in analyzing the energy and work change related to the process.

Adiabatic process: This is a thermodynamic process in which there is no heat transfer into or out of the system. For this process, change in quantity of heat is zero (i.e. $\Delta Q \equiv 0$ during this process).

Isochoric process: This is a thermodynamic process that occurs at constant volume (i.e. $\Delta V \equiv 0$ during this process). This implies that during this process no work is done on or by the system.

Isobaric process: This is a thermodynamic process that occurs at constant pressure (i.e. $\Delta P \equiv 0$ during this process).

Isothermal process: This is a thermodynamic process that takes place at constant temperature (i.e. $\Delta T \equiv 0$ during this process).

PROPERTY OF SYSTEM: -

Any characteristic (quality/feature) of a system is called property.

Intensive property: - Intensive properties are those that are independent of the mass of a system such as temperature, pressure etc.

Extensive property: - Extensive properties are those that are dependent on mass of a system.

An easy way to determine whether a property is intensive or extensive is to divide the system into two equal parts with an imaginary partition. Each part will have the same value of intensive property as the original system but half the value of extensive properties.

Demonstration: - As an illustration of these two categories of thermodynamic properties, assuming you cut a hot bar of metal of uniform temperature T into two equal parts. Each half will still have almost the same temperature T. This clearly shows that temperature is independent of mass of the 'system,' (i.e. metal bar). But on the other hand, the volume of each part indicates that volume of a system is dependent of mass (volume is extensive property).

Specific property: - It is defined as extensive property per unit mass of system. It is a special case of an intensive property. For example, Specific volume V_s is

$$V_s = \frac{V}{M};$$

CONCEPT OF CONTINUUM: -

The atomic nature of a substance which can be viewed as continuous that is homogeneous matter with no holes can be considered to follow continuum idealization. This idealization allows us to treat properties as point function and to assume the properties vary continually in space with no discontinuities. This is valid as long as the size of system we deal with is large relative to the space between the molecules.

SATE AND EQUILIBRIUM: -

State: - A set of properties that completely describes condition of a system is known as state of the system. At a given state, all properties of a system have fixed values. In other words, it is the condition of a system as defined by the values of all its properties. It gives a complete description of the system.

Equilibrium: - The word equilibrium implies a state of balance. In an equilibrium state, there presents no driving force within the system. A system is said to be in thermodynamic equilibrium with its surrounding or with another system if and only if the system is in thermal equilibrium, in chemical equilibrium and in mechanical equilibrium with the surrounding or with another system. If any one of the above conditions is not fulfilled, the system is not in thermodynamic equilibrium.

Thermal equilibrium: - A system is in thermal equilibrium if the temperature is same throughout the entire system. That is, the system involves no temperature difference, which is the driving force for heat flow.

Mechanical equilibrium: - A system is in mechanical equilibrium if there is no change in pressure at any point of the system with time. However, the pressure may vary within the system with elevation as a result of gravitational effects.

Phase equilibrium: - A system is in phase equilibrium when the mass of each phase reaches an equilibrium level and stays there.

Chemical equilibrium: - A system is in chemical equilibrium if its chemical composition does not change with time, i.e., no chemical reaction occur.

STATE POSTULATE: -

The number of properties required to fix the state of a system is given by state postulate: "the state of a simple compressible system is completely specified by two independent, intensive properties". Also, remember that temperature and pressure are independent property for single phase system but are dependent for multiphase system.

PROCESSES AND CYCLES: -

Process: - Any change that a system undergoes from one equilibrium state to another is called a process and the series of states through which a system passes during a process is called path of the process.

Reversible Process: A reversible process can be defined as one in which direction can be reversed by an infinitesimal change in some properties of the system.

Irreversible Process: An irreversible process can be defined as one in which direction cannot be reversed by an infinitesimal change in some properties of the system.

Quasi equilibrium process: - When process proceeds in such a manner that the system remains infinitesimally close to an equilibrium state at all times, it is called a quasi-static or quasi equilibrium process. A quasi equilibrium process can be viewed as a sufficiently slow process that allows the system to adjust itself internally so that properties in one part of the system do not change any faster than those at other parts.

Quasi-Equilibrium Process

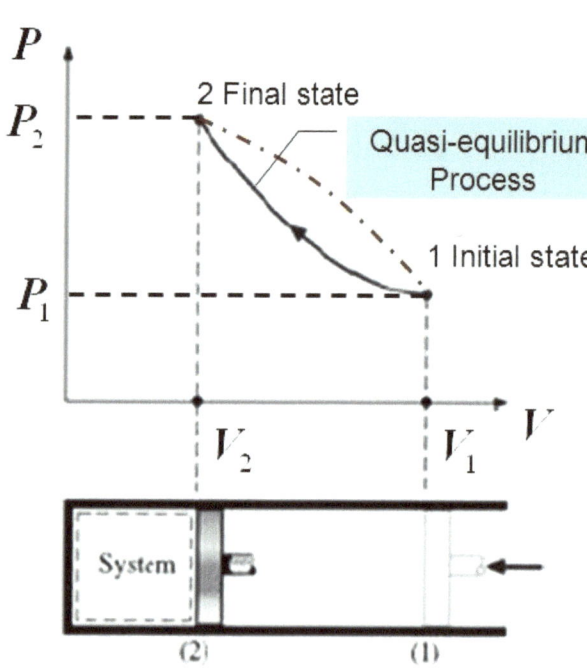

System remains infinitesimally close to an equilibrium state at all times.

Process occurs slow enough to keep the properties identical throughout the system.

The interval of process change takes much longer time than system spontaneously adjusts to a new state after the previous equilibrium state was destroyed.

To connect states with line must be QE process.

<u>Reason for assuming quasi equilibrium process: -</u> They are easy to analyze. Work producing devices produces more work when operated on quasi equilibrium process.

Non-quasi-static Process: This is a process that is carried out in such a way that at every instant, there is finite departure of the system from an equilibrium state.

Cycle: - A system is said to have undergone a cycle if it returns to its initial state at the end of the process.

TEMPERATURE AND THE ZEROTH LAW OF THERMODYNAMICS: -

Temperature may define as a measure of hotness or coldness. Temperature is one of the seven S.I. base quantities and it is measure in Kelvin (K). Temperature of a body has no upper limit but it has lower limit (absolute zero or zero Kelvin).

Heat (Q) is a form of energy that is transferred from one part of a substance to another or from one body to another by virtue of a difference in temperature (i.e. temperature gradient). The unit of heat is in Joules (J).

Sign convention of heat (Q): -

Q is positive when there is a flow of heat into the system.
Q is negative when there is a flow of heat out of the system.

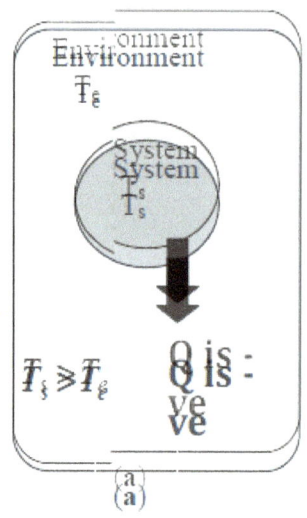
$T_s > T_e$ Q is -ve
(a)

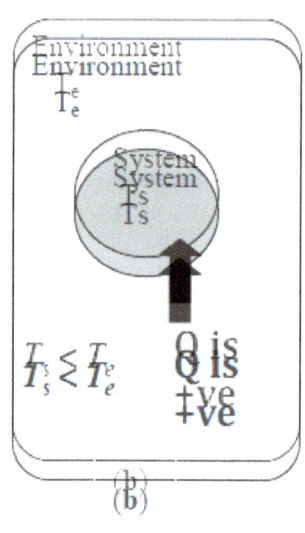
$T_s < T_e$ Q is +ve
(b)

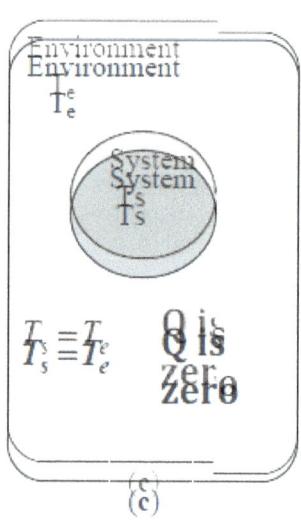
$T_s \equiv T_e$ Q is zero
(c)

Process (a) denotes heat is rejected from the system.

Process (b) denotes heat is added to the system.

Process (c) denotes no heat is added.

ZEROTH LAW OF THERMODYNAMICS: -

It states that if two bodies are in thermal equilibrium with a third body, then, they all are in thermal equilibrium with each other.

Demonstration: -

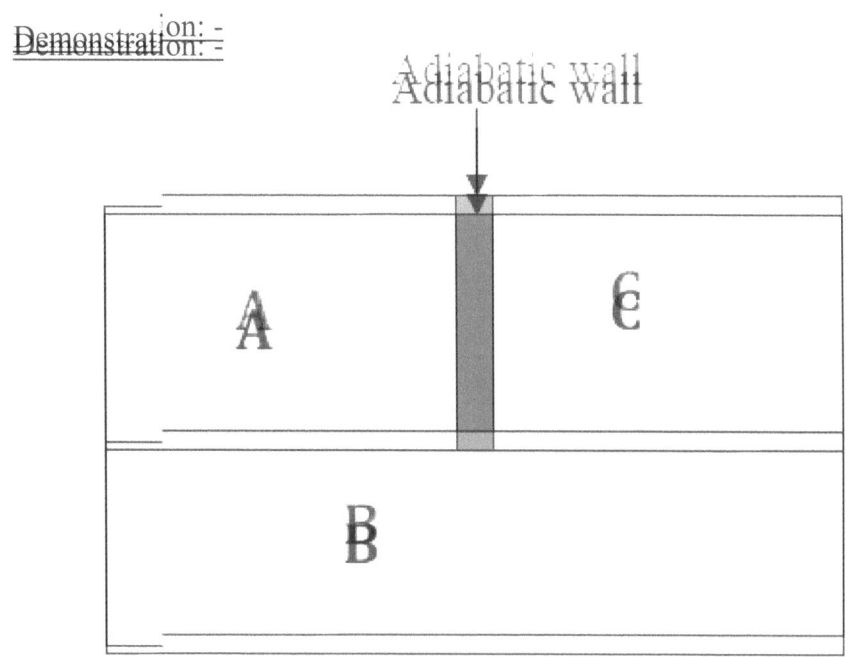

Suppose body B is in thermal contact with bodies A and C but bodies A and C are not in thermal contact. If B is in thermal equilibrium with A and C; then A and C are in thermal equilibrium.

Explanation of Zeroth Law: -

Let us say T_A, T_B and T_C are the temperatures of A, B and C respectively.

A and B are in thermal equilibrium. That is, $T_A = T_B$.

C and B are in thermal equilibrium. That is, $T_B = T_C$.

Then using Zeroth law of thermodynamics, $T_A = T_C$.

Note: - All temperature measurements are based on this Law.

TEMPERATURE SCALES: -

Available temperature scales are; the Celsius scale (also known as the Centigrade scale), the Fahrenheit scale, the Kelvin scale, the Rankine scale, and the international thermodynamic temperature scale.

Conversion between Temperature Scales: -

Kelvin Scale to Celsius Temperature Scale: -

The relation between Kelvin scale and Celsius scale is

$$T_c = (T - 273.15)\,°C$$

Where T_C is the temperature in degree centigrade, and T is the temperature in Kelvin.

Fahrenheit Temperature scale: -

The relation between Celsius scale and Fahrenheit scale is

$$T_F = \frac{9}{5}T_c - 32$$

Where T_C is the temperature in degree centigrade, and T_F is the temperature in Fahrenheit.

TYPES OF THERMOMETERS: -

Thermometer is named after the thermometric property used for the construction. For example, thermometric property of the mercury-in glass thermometer is the length of mercury column, while in the constant-volume gas thermometer the thermometric property is the pressure of the gas.

Thermocouple Thermometer: -

It has been observed that when two dissimilar metals are joined together to make two junctions, an electromotive force (emf) will flow in the circuit. This emf can be measured using a voltmeter and its value depends on the temperature difference between the junctions. The arrangement is called thermocouple and the observation is known as **Seebeck effect**. Thermocouple thermometer is based on the **Seebeck effect**.

Thermometric property: - emf generated when two junctions made from two different metals are maintained at different temperature.

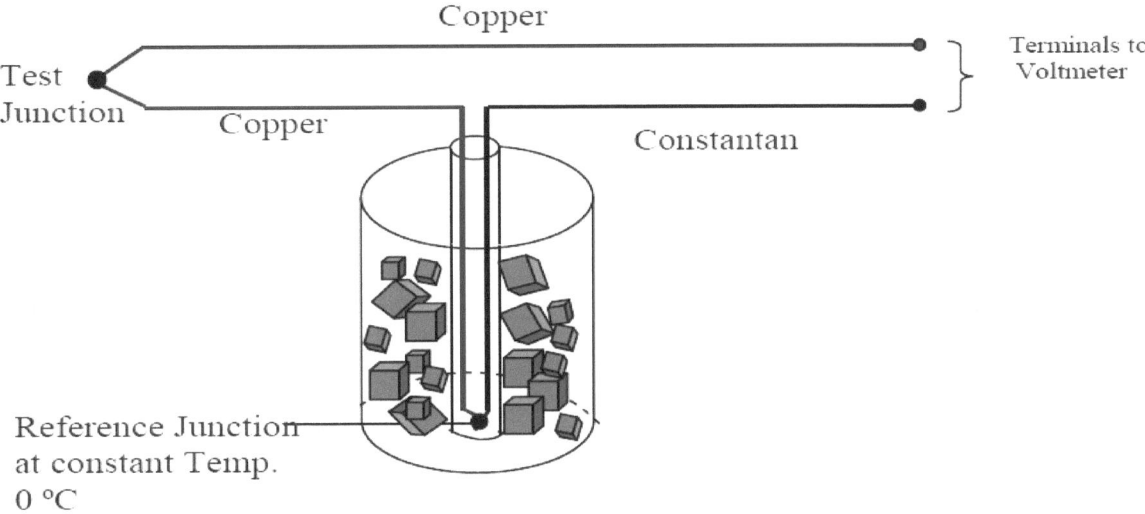

The diagram shows the arrangement for the copper constantan thermocouple thermometer. The test junction is placed on the body or inside the system whose temperature is to be measured, while the reference junction is maintained at constant temperature at 0 °C. The potentiometer is connected to the terminals to the voltmeter. The relationship between the *emf* and temperature is

$$emf = a + bT + cT^2 + dT^3$$

Where **a, b, c,** and **d** are constant and they are different for each thermocouple.

Thermocouple thermometer is used extensively in scientific laboratories.

Resistance Thermometer: -

Electrical conductivity of a metal depends on the movement of electrons through its crystal lattice. The electrical resistance of a conductor, due to thermal excitation, varies with temperature. This forms the basic principle of operation of resistance thermometer Resistance thermometer. Therefore, uses the variation in electrical conductivity of a conductor to indicate temperature.

Demonstration: -

Figure shows a generalized form of a resistance thermometer. In the diagram, **RT** is the resistance element which could be any conductor (e.g. platinum). This is usually wound round a frame constructed so as to avoid excessive strains when the wire contracts upon cooling. **S** is the power supply and the purpose is to maintain a known constant current in the thermometer while measuring the potential difference with the aid of a bridge output (usually a sensitive potentiometer).

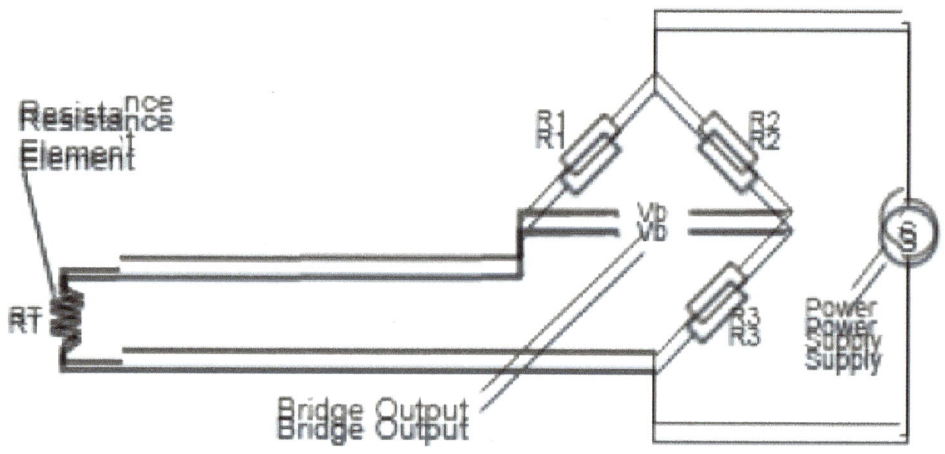

The relationship between the temperature and the electrical resistance is usually non-linear and described by a higher order polynomial:

$$R(T) \equiv R_0(1 + AT + BT^2 + CT^3 + \ldots)$$

Where T is the Celsius temperature, R_o is the nominal resistance at a specified temperature, and A, B, C are constants. The number of higher order terms considered is a function of the required accuracy of measurement. The constants (i.e. A, B and C etc.) depend on the conductor material used and basically define the temperature-resistance relationship. The value R_0 is referred to as nominal value or nominal resistance and is the resistance at $0°C$. Material most commonly used for resistance thermometers are Platinum, Copper and Nickel. However, Platinum is the most dominant material internationally.

The range of measurement of resistance thermometer depends on the choice of conductor used. Platinum resistance has a very accurate measurement within -253 to 1200 °C.

Constant-Volume Gas Thermometer: -

Constant-Volume gas thermometer is normally referred to as the standard thermometer and is mostly used to calibrate other thermometers. The working principle is based on the pressure of a gas in a fixed volume. Figure shows an example of a constant-volume gas thermometer. It consists of a gas-filled bulb connected by a tube to a mercury manometer. By raising and lowering reservoir R, the mercury level on the left can always be brought to the zero of the scale to keep the gas volume constant (Note that variation in the gas volume can affect temperature measurement).

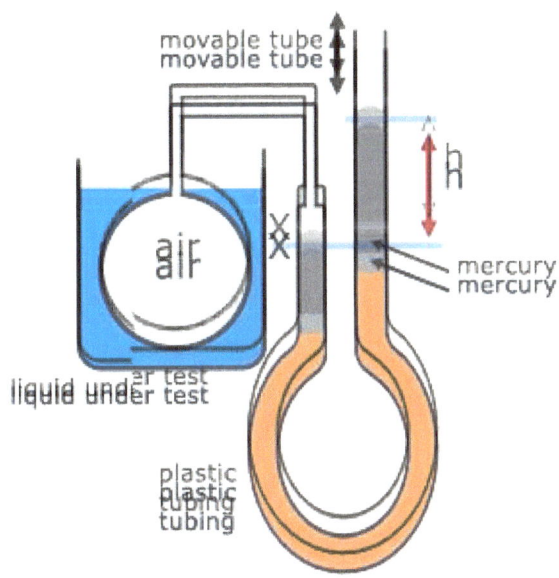

The basic equation is

$$P = P_0 = \rho g h$$

Where P_0 is the atmospheric pressure, r is the density of the mercury in the manometer, g is the acceleration due to gravity, and h is the measured difference between the mercury level in the two arms of the tube.

Chapter – 2

ENERGY AND ITS FORMS

"Energy conservation is the conservation of the quality of energy, not the quantity". Work, which is the highest quality of energy, can always be converted to an equal amount of thermal energy but a small fraction of thermal energy, can be converted back to work.

Energy can exist in numerous forms such as thermal, mechanical, kinetic, potential, electric, magnetic, chemical and nuclear and their sum constitutes the total energy of a system.

Thermodynamics provides no information about the absolute value of the total energy. It deals only with the change of the total energy. Thus the total energy of a system can be assigned a value of zero at some convenient reference point. The change in total energy of a system is independent of the reference point selected.

The macroscopic form of energy are those a system possesses as a whole with respect to some outside reference frame such as kinetic and potential energies. The microscopic forms of energy are those related to molecular structure of a system and the degree of molecular activity and they are independent of outside reference frames. The sum of all microscopic form of energy is called internal energy of a system and is denoted by U.

The kinetic energy of a system is expressed as

K.E = $mv^2/2$ (kJ) where m – mass, v – velocity,

Similarly, potential energy of a system is expressed as

P.E = mgh (kJ) where h – elevation of a system

Therefore, in absence of electric, magnetic, surface tension effects, the total energy of a system consist of kinetic, potential and internal energies as

E = U + K.E + P.E = U + $mv^2/2$ + mgh (kJ)

An energy interaction is heat transfer if its driving force is a temperature difference otherwise it is work. A control volume can also exchange energy via mass transfer since any time mass is transferred into or out of a system, the energy content of the mass is also transferred with it. The sensible and latent energies are commonly referred as thermal energy to prevent any confusion with heat transfer.

Notes on Heat: -

All temperature changes need not be due to heat alone example: - friction.

All heat interaction need not result in changes in temperature, example: - evaporation and condensation.

METHODS OF HEAT TRANSFER: -

The transfer of heat from one part of a system to another or to another system by virtue of a temperature difference can only be one or more of the three processes namely; conduction, convection, and radiation.

Conduction: -

This is the process of heat transfer whereby heat energy is transferred directly through a material without any bulk movement of the material.

Why do conductors conduct?

The behavior of conductors in terms of thermal conductivity can be explained with two mechanisms namely; collision mechanism and free electrons in metals.

Molecular collision: Atoms and molecules in a hot part of the material vibrate or move with greater velocity (i.e. higher kinetic energy) than those at the colder part. By means of collisions, the more energetic molecules pass on a portion of their energy to their less energetic neighbors. As the more energetic molecules collide with their less energetic neighbors they transfer some of their energy to the neighbors. The collision mechanism does not depend on bulk movement of the material.

Free electrons in metals: Good conductors of thermal energy, like metals, have pool of electrons that are more or less free to wander through the volume of the metal. These free electrons are capable of transporting energy round the whole volume of conductors. Free electrons are also responsible for the excellent electrical conductivity in metals.

Convection: -

Convection is the process in which heat energy is transferred from place to place by the bulk movement of a fluid. A good example of this process is convection current in liquid.

Radiation: -

This is a process in which energy is transferred by means of electromagnetic waves. A good example of this is the solar radiation from the sun traveling in all directions in space. All bodies, hot or cold, continuously radiate energy in form of electromagnetic waves. But the amount of this radiation is proportional to the temperature of the body and the nature of its surface.

HEAT AND WORK: -

Heat is defined as the form of energy that is transferred between two systems by virtue of a temperature difference. That means heat is energy in transition that is heat means heat transfer. A process during which there is no heat transfer is called an adiabatic process. Although, no heat transfer during an adiabatic process, the energy content and thus, the temperature of a system can still be changed by other means such as work.

If the energy crossing the boundary of a closed system is not heat, it must be work. That means if energy interaction is not caused by temperature difference between systems and surrounding is work. Heat and work are directional quantities and thus, they require the specification of both the magnitude and direction. A quantity that is transferred to or from a system during an interaction is not a property since the amount of such a quantity depends on more than just the state of the system.

Sign convention: - Heat transfer to a system and workdone by a system is positive. Heat transfer from a system and workdone on a system is negative.

Important points to remember: - Both heat and work are recognized at the boundaries of a system as they cross the boundaries. That is, heat and work are boundary phenomena. A system can possess energy but not work or heat. Unlike properties, heat and work has no meaning at a state since they are both path dependent.

Imagine a hydrostatic system contained in a cylinder with a movable piston like the one shown in figure below. From the diagram, suppose an external force F acted in the direction showed moving the piston from initial point 1 to final point 2 through a distance dx. Suppose that the cylinder has a cross section area A; that the pressure exerted on the system at the piston face is P; and that the force is PA. The system also exerts an opposing force on the piston. The work done dW on the system in the process described above is

$$dW = PAdx$$

but $$Adx = dV$$

Therefore $$dW = -PdV$$

The negative sign in the last equation indicate negative change in volume (i.e. a decrease in volume).

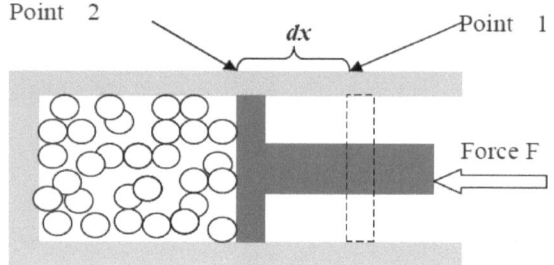

In a finite quasi-static process in which the volume changes from V_i to V_f, the work done is

$$W = -\int_{V_i}^{V_f} P\,dV$$

Work in Quasi-Static Process: -

For a quasi-static isothermal expansion of compression of an ideal gas

$$W = -\int_{V_i}^{V_f} P\,dV$$

But an ideal gas is the one which equation of state is **PV = nRT**, where **n** and **R** are constant. Replace **P** with **nRT/V** in above equation,

$$W = -\int_{V_i}^{V_f} \frac{nRT}{V}\,dV$$

And since **T** is constant for isothermal process,

$$W = -nRT \int_{V_i}^{V_f} \frac{dV}{V}$$

on integration, we get

$$W = -nRT \ln \frac{V_f}{V_i}$$

WORK AND INTERNAL ENERGY: -

When an adiabatic work (dW_{ad}) is done on or by the system, the internal energy of the system changes. The change in internal energy (ΔU) is equal to the adiabatic work done.

$$dU = dW_{ad}$$

If a system changes from state 1 to state 2 by doing adiabatic work, and if the states are differed by a finite amount, then

$$\int_{U_1}^{U_2} dU = U_2 - U_1 = \int_1^2 dW_{ad} = -W_{ad}$$

Suppose the work done is mechanical work (i.e. mechanical adiabatic work), then

$$U_2 - U_1 = -\int_1^2 PdV$$

OTHERS FORMS OF WORK: -

Stretching of a wire: -

Let a wire be stretched by dL due to the application of a force F; then, Work is done on the system. Therefore dW= - FdL.

Electrical Energy: -

Flowing in or out is always deemed to be work dW= - EdC= - EIdt.

Work due to stretching of a liquid film due to surface tension: -

Let us say a soap film is stretched through an area dA; dW= - σdA where σ is the surface tension.

Chapter – 3

FIRST LAW OF THERMODYNAMICS

First state of thermodynamics states that energy can neither be created nor be destroyed during a process but it can be converted from one form to another. In mathematical term, it can be defined as

$\Delta E = \Delta Q + \Delta W$; where ΔE = change in total energy of system

ΔQ ≡ change in heat energy of system

ΔW ≡ change in workdone

If the system is adiabatic, then the net workdone is same regardless of the nature of closed system and the details of the process.

That is if $\Delta Q = 0$; then, this implies to $\Delta E = \Delta W$.

The **essence of first law** is the existence of the property **total energy**. In simple words, the first law is an expression of the conservation of energy principle and it asserts that energy is a thermodynamic property.

Demonstration: - suppose we compressed a insulted piston cylinder container filled with air, then, the temperature of air rises. This is because energy is transferred to air in the form of boundary work.

ENERGY BALANCE EQUATION: -

It can be defined as "the net change in the total energy of the system during a process is equal to the difference between the total energy entering and total energy leaving the system during that process".

That is, $E_{in} - E_{out} = \Delta E$

Since energy is a property and the value of property does not change unless the state of the system changes. Therefore, the energy change of system is zero.

Also, $\Delta E = \Delta U + \Delta K.E + \Delta P.E$

For stationary system, $\Delta K.E = 0, \Delta P.E = 0$, this implies $\Delta E = \Delta U$.

Also,

$E_{in} - E_{out}$ means net energy transfer by heat, work and work.

ΔE means change in internal, kinetic and potential energies.

For a closed system undergoing cycle, initial and final states are identical and thus,

$\Delta E = E_{final} - E_{initial} = 0$

This implies, $W_{net,\,out} = Q_{net,\,in}$

Chapter – 4

PURE SUBSTANCES

A substance that has a fixed chemical composition throughout is called a pure substance. Water, carbon dioxide and nitrogen are example of pure substance. A pure substance does not have to be of a single chemical elements or compounds, if mixture is homogeneous, it is a pure substance. Like, air is a mixture of several gases, but it is often considered to be a pure substance because it has a uniform chemical composition. A mixture of two or more phases of a pure substance is still a pure substance as long as the chemical composition of all phases is the same.

Demonstration: - A mixture of liquid air and gaseous air is not a pure substance since the composition of liquid air is different from the composition of gaseous air and thus the mixture is no longer chemically homogeneous. This is due to different components in air condensing at different temperature at a specified pressure.

PHASES OF A PURE SUBSTANCE: -

A phase is identified as having a distinct molecular arrangement that is homogeneous throughout and separated from others by easily identifiable boundary surfaces. Intermolecular bonds are strongest in solids and weakest in gases. Simple reason behind this is that molecules in solids are closely packed together, whereas in gases they are separated by relatively large distances. The attractive forces between molecules turn to repulsive forces as the distance between the molecules approaches zero, thus preventing the molecules from piling up on top of each other.

The molecules in a solid can't move relative to each other, they continually oscillate about their equilibrium position. The velocity of molecule during these oscillations depends on the temperature. At sufficiently high temperature, the velocity and thus the momentum of the molecules may reach a point where the intermolecular forces are partially overcome and groups of molecules break away this is the beginning of the melting process.

The molecular spacing in the liquid phase is not much different from that of the solid phase, except the molecules are no longer at fixed position relative to each other and they can rotate and translate freely. While molecules in the gas phase are at considerably possess higher energy level than they are in the liquid or solid phases. Therefore, the gas must release a large amount of its energy before it can condense or freeze.

PHASE CHANGE PROCESSES: -

Compressed liquid and saturated liquid: - consider a piston cylinder device containing liquid water at normal condition that is at 25°C and 1 atm pressure. At this condition, water exist in the liquid phase and it is called a compressed liquid or a sub cooled liquid meaning that it is not about to vaporize. Keeping the pressure constant, heat is now transferred to water until it reaches 100°C. At this point, water is still a liquid, but any heat addition will cause some of liquid to vaporize. That is, phase change process from liquid to vapor is about to take place. A liquid that is about to vaporize is called a saturated liquid.

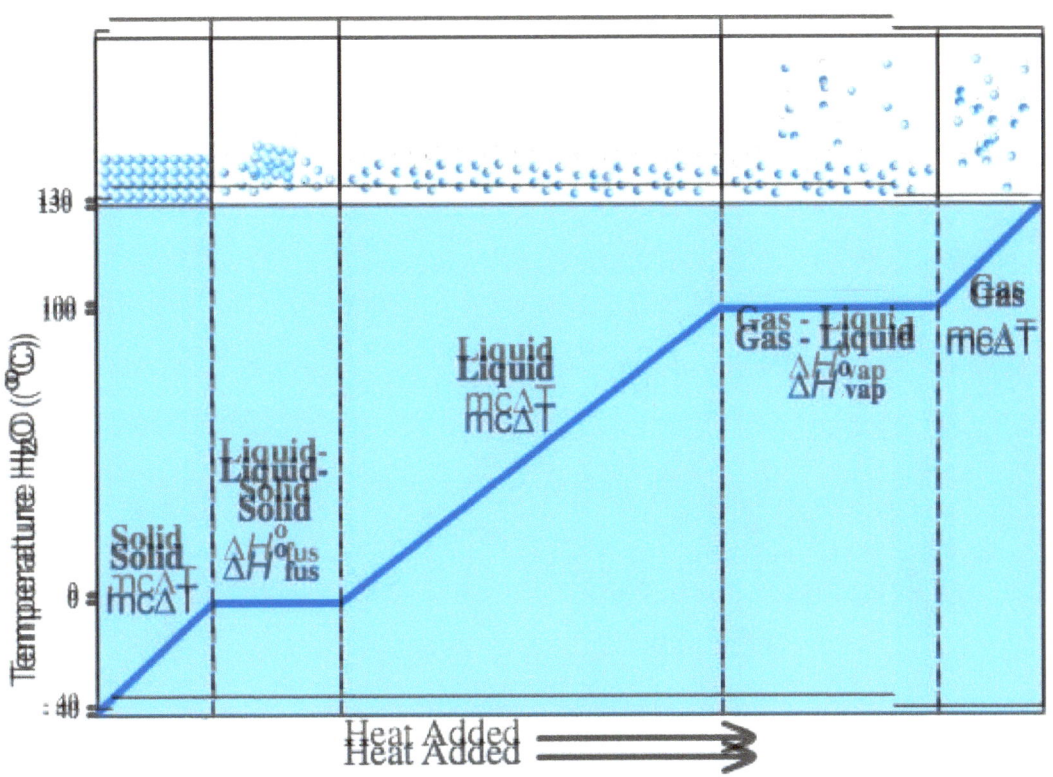

Saturated vapor and superheated vapor: - the temperature will remain constant during the entire phase change process if pressure is held constant. When liquid water is completely vaporized, the entire cylinder is filled with vapor that is on the verge of liquid phase. Any heat loss from this vapor will cause some of vapor to condense. A vapor that is about to condense is called a saturated vapor. A substance that is in between the phase change process is referred to as a saturated liquid vapor mixture since the liquid and vapor phases coexist in equilibrium at these states. Once the phase change process is completed, further transfer of heat results in an increase in both the temperature and the specific volume. At this time, a vapor is called as superheated vapor since it is not about to condensate.

SATURATION TEMPERATURE AND SATURATION PRESSURE: -

The temperature at which water starts boiling depends on the pressure; therefore if the pressure is fixed, so is the boiling temperature. At a given pressure, the temperature at which a pure substance changes phase is called saturation temperature (T_{sat}). Similarly, at a given temperature, the pressure at which a pure substance changes phase is called the saturation pressure (P_{sat}).

The amount of energy absorbed or released during a phase change process is called latent heat. The amount of energy absorbed during melting is called latent heat of fusion and is equivalent to the amount of energy released during freezing. Similarly, amount of energy absorbed during vaporization is called latent heat of vaporization and is equivalent to the energy released during condensation. The magnitude of latent heat depends in the temperature or pressure at which the phase change occurs. During a phase change process, pressure and temperature are obviously dependent properties and therefore, $T_{sat} = f(P_{sat})$.

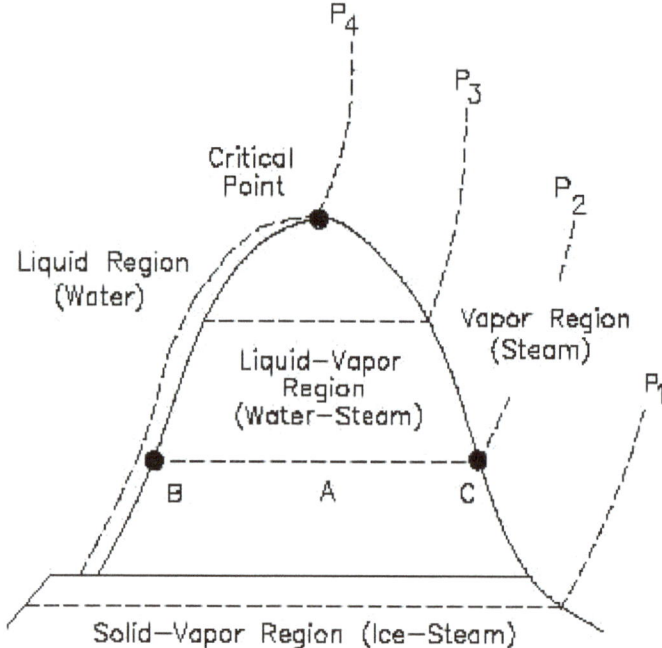

In previously explain assumption; now let's increase its pressure. Then, the first thing which we observe is that water starts boiling at a much higher temperature and secondly, the specific volume of the saturated liquid is larger and the specific volume of the saturated vapor is smaller than the corresponding values at 1 atm pressure. That is, the horizontal line that connects the saturated liquid and saturated vapor states is much shorter. As the pressure is increased further, this saturation line continues to shrink and it becomes a point when the pressure reaches 22.06 MPa (for water).

This point is called critical point and it is defined as the point at which the saturated liquid and saturated vapor states are identical. The temperature, pressure and specific volume of a substance at the critical point are called critical temperature (T_{cr}), critical pressure (P_{cr}), critical specific volume (v_{cr}).

At pressure above the critical pressure, there is not a distinct phase change process. Instead, the specific volume of the substance continually increases and at all times there is only one phase present and it's obvious that it resembles a vapor, but we can never specify when the change has occurred. Above the critical state, there is no line that separates the compressed liquid region and the superheated vapor region.

PROPERTY TABLE:-

For each substance, a separate table is prepared for each region of interest such as superheated vapor, compressed liquid and saturated region.

Property Table
Saturated water – Temperature table

Temp T_{sat}, °C	Specific internal energy, kJ/kg			Specific enthalpy, kJ/kg		
	u_f, kJ/kg	u_{fg}, kJ/kg	u_g, kJ/kg	h_f, kJ/kg	h_{fg}, kJ/kg	h_g, kJ/kg
10	42.00	2347.2	2389.2	42.01	2477.7	2519.8
50	209.32	2234.2	2443.5	209.33	2382.7	2592.1
100	418.94	2087.6	2506.5	419.04	2257.0	2676.1
300	1332.0	1231.0	2563.0	1332.0	1940.7	2793.2
374.14	2029.6	0	2029.6	2099.3	0	2099.3

The subscript 'f' is used to denote property of a saturated liquid and the subscript 'g' to denote the property of saturated vapor and 'fg' denotes the difference between the saturated vapor and saturated liquid values of the same property. The quantity h_{fg} is called enthalpy of vaporization or latent heat of vaporization. It represents the amount of energy needed to vaporize a unit mass of saturated liquid at a given temperature or pressure. It decreases as temperature or pressure increases and becomes zero at critical point.

SATURATED LIQUID VAPOR MIXTURE: -

During vaporization process, a mixture of saturated liquid and saturated vapor exist, to analyze this mixture properly, we need to know the proportion of the liquid and vapor phase in the mixture. This is by defining a new property called the quality, x as the ratio of mass of vapor to the total mass of the mixture.

$x \equiv$ mass of vapor / total mass of mixture

$M_T \equiv M_{liquid} + M_{vapor} = M_f + M_g$

Quality has significance for saturated mixtures only. It has no meaning in the compressed liquid or superheated vapor regions. It's value is between 0 & 1.

The properties of saturated liquid are the same whether it exist alone or in a mixture with saturated vapor. During the vaporization process, only the amount of saturated liquid changes not its properties. The same can be said about saturated vapor.

TYPES OF PHASE TRANSITION: -

The three types of phase transitions are: first order, second order and lambda phase transitions.

First Order Phase Transition: -

The phase transitions that we are familiar with i.e. sublimation, vaporization and fusion are called first order phase transition. They are called first order because the first order derivatives of the Gibbs function are finite.

Therefore, for a first order phase transition:

1. There are changes in entropy and volume, and
2. The first-order derivatives of the Gibbs function change discontinuously.
3. The specific heat capacity at constant pressure is infinite, this is because temperature is constant during phase change.

$$(C_p = T \left.\frac{\partial S}{\partial T}\right|_P)$$

Second Order Phase Transition: -

This is a phase transition in which the second derivates of Gibbs are finite.

For order phase transition,

1. T, P, G, S, and V (also H, U, and F) remain unchanged

2. C_P, β, and κ (i.e. from second derivatives of G) are finite.

The only example of second order phase transition is the transition for the superconductor from superconducting to the normal state in zero magnetic fields.

Lambda phase transition: -

For the lambda phase transition: -

1. T, P, and G remain constant;
2. S and V (also U, H, and F) remain constant; and
3. C_P, β, and κ are infinite

The most interesting example of lambda transition is the transition from ordinary liquid helium to super fluid helium at a temperature and corresponding pressure known as a lambda point.

IDEAL GAS EQUATION:-

Any equation that relates the pressure, temperature and specific volume of a substance is called equation of state. The vapor phase of a substance is called gas when it is above the critical temperature.

Ideal gas equation predicts P-v-T behavior of a gas quite accurately within some properly selected region.

$Pv = RT$, where R = gas constant; P = absolute pressure

v = specific volume; T = absolute temperature

The gas constant is different for each gas and is determined from $R = R_u / M$ (kJ/kg.K)

Where R_u = universal gas constant; M = molar mass

The properties of an ideal gas at two different states are related to each other by

$P_1 V_1 / T_1 = P_2 V_2 / T_2$.

An ideal gas is an imaginary substance that obeys ideal gas equation. At low pressure and high temperature, the density of gas decreases and gas behaves as an ideal gas under these conditions.

In air conditioning applications, water vapor in air can be treated as an ideal gas with essentially no error since the pressure of vapor is very low. In steam power plant applications, however the pressure involved are usually very high. Therefore, ideal gas relation should not be used.

REAL GAS: -

A real gas obviously does not obey the perfect gas equation because, the molecules have a finite size (however small it may be) and they do exert forces among each other. Equations derived to describe the real gases is the Vander Waal's equation

$$(P+a/v^2)(v-b)=RT$$

Constant *a* takes care of attractive forces; *b* the finite volume of the molecule.

For a real gas $PV \neq RT$ and the quantity $PV/RT = z$ and is called the "COMPRESSIBILITY".

Also, for a perfect gas $z = 1$.

Chapter – 5

CONCEPT OF BOUNDARY WORK

The work associated with a moving boundary is called boundary work. The moving boundary work associated with real engines or compressors cannot be determined exactly from a thermodynamics analysis alone because the piston usually moves at very high speeds, making it difficult for the gas inside to maintain equilibrium. Then the states through which the system passes during the process cannot be specified and no process path can be drawn. Work, being a path function, can't be determined analytically without knowledge of path. Therefore, the boundary work in real engines or compressors are determined by direct measurement.

Consider a gas enclosed in piston cylinder device. The initial pressure of the gas is P, the total volume is V and the cross sectional area of the piston is A. If the piston is allowed to move a distance 'ds' in a quasi equilibrium manner, the differential workdone during this process is

δW_b = F x (ds) = P x A x (ds) = P x dv

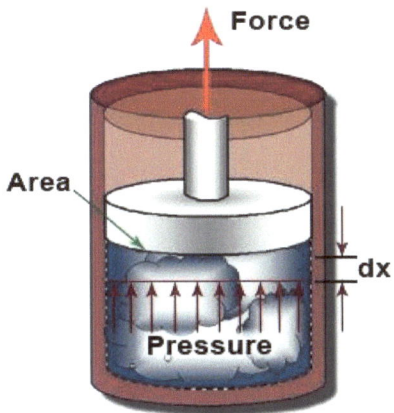

That is, the boundary work in the differential form is equal to the product of absolute pressure P and the differential change in the volume 'dv' of the system. Also, because of this expression, moving boundary work is sometimes called the P x dv.

However, P is the absolute pressure, which is always positive. But, the volume change 'dv' is positive during an expansion process and negative during a compression process. A negative result indicates boundary work input that is compression.

The total boundary work done during the entire process as the piston moves is obtained by adding all the differential works from initial state to the final state;

$W_b = \int_1^2 P dV$ (kJ)

This integral can be evaluated only if we know the functional relationship between P and V during the process. That is, P = f(v) should be available. Note that P = f(v) is simply the equation of the process path on the P-v diagram.

If work were not a path function, no cyclic devices (car engines, power plants) could operate as work producing devices. The work produced by these devices during one part of the cycle would have to be consumed during another part and there would be no net work output.

Polytropic process: -

During actual expansion and compression process of gases, pressure and volume are often related to $PV^n = C$; where C and n are constants. A process of this kind is known as Polytropic process. The pressure for a Polytropic process can be expressed as $P = CV^{-n}$;

$$W_b \equiv \int_1^2 P dV = \int_1^2 CV^{-n} dV = (P_2 V_2 - P_1 V_1)/(1-n)$$

SPECIFIC HEAT:-

The specific heat is defined as the energy required to raise the temperature of a unit mass of a substance by one degree. Specific heat at constant volume (C_v) can be defined as the energy required to raise the temperature of the unit mass of a substance by one degree as the volume is maintained constant. The energy required to do the same as the pressure is maintained constant is the specific heat at constant pressure (C_p).

Specific heat at constant pressure is always greater than specific heat at constant volume because at constant pressure the system is allowed to expand and the energy for this expansion work must also be supplied to the system.

An expression for the specific heat at constant volume can be obtained by considering constant volume process, thus no expansion or compression work is involved. It yields

$e_{in} = e_{out} \equiv du$;

Hence from definition of C_v, the energy must be equal to $C_v dT$, where dT is the differential change in temperature. Thus,

$C_v dT \equiv du$; at constant volume

$C_v \equiv (du/dT)_v$

Similarly, an expression for specific heat at constant pressure C_p can be obtained by considering a constant pressure expansion or compression process. It yields

$C_p \equiv (dh/dT)_p$

Like any other property, the specific heat of a substance depends on the state that, in general, is specified by two independent, intensive properties. That is, the energy required to raise the temperature of a substance by one degree is different at different temperature and pressure.

These above equations are property relation and as such are independent of the type of processes that is they are valid for any substance undergoing any process. Also C_v is a measure of variation of internal energy of substance with temperature and C_p is a measure of variation of enthalpy of a substance with temperature.

For an ideal gas, the internal energy is a function of temperature only that is $u = u(T)$.

Also, $h = u + Pv$ and $Pv = RT$

This implies $h = u + RT$ (since R is constant and $u = u(T)$, it follows that the enthalpy of an ideal gas is also a function of temperature only).

$h = f(T)$

Since u and h depend only on temperature for an ideal gas, the specific heat C_v and C_p also depend on temperature only. Therefore, at a given temperature u, h, C_v and C_p of an ideal gas have a fixed values regardless of the specific volume or pressure.

For ideal gas, $du = C_v(T)dT$ and $dh = C_p(T)dT$

SPECIFIC HEAT RELATION OF IDEAL GASES: -

We know that $h = u + RT$

On differentiating, $dh = du + RdT$ => $\mathbf{C_p = C_v + R}$; ($dh = C_p\, dT$ and $du = C_v\, dT$)
Also, the specific heat ratio also varies with temperature but this variation is very mild. For mono atomic gases, k = 1.667 and for diatomic gases, k = 1.4 at room temperature.

INTERNAL ENERGY, ENTHALPY AND SPECIFIC HEATS OF SOLID AND LIQUID:

A substance whose specific volume or density is constant is called an incompressible substance. The constant volume assumption should be taken to imply that the energy associated with volume change is negligible with other forms of energy.

That is for solid and liquid, $C_p = C_v = C$.

Therefore, specific heat of incompressible substance depends on temperature only.

That is $\quad du = C_v\, dT = C(T)\, dT \quad \text{or} \quad \Delta u = u_2 - u_1 = \int_1^2 C(T)\, dT$

Similarly, $\quad dh = du + vdP + Pdv \Rightarrow dh = du + vdP \text{ (kJ/kg)}$

CONCEPT OF CONSERVATION OF MASS: -

Mass like energy is a conserved property and it cannot be created or destroyed during a process. However, mass m and energy E can be converted to each other. A fluid usually flows into or out of a control volume through pipes or ducts. The differential mass flow rate of fluid flowing across a small area element dA_c itself, the fluid density ρ, and the component of the flow velocity normal to dA_c, which we denote as v_n, is $\delta m = \rho \times v_n \times dA_c$

Note that both δ and d are used to indicate differential quantities, but δ is typically used for quantities (such as heat, work and mass transfer) that are path function and have exact differential while 'd' is used for quantities (such as properties) that are point function and have exact differentials.

The mass flow rate through the entire cross sectional area of a pipe or duct is obtained by integration $\dot{m} = \int_{AC} \delta m = \int_{AC} (\rho \times v_n \times dA_c)$ (kg/s)

In general, compressible flow both ρ and v_n vary across the pipe. In many practical applications, however the density is essentially uniform over the pipe cross section and we can take 'ρ' outside the integral. Velocity is never uniform over a cross section of a pipe because of the fluid sticking to the surface and thus having zero velocity at wall (no – slip condition). Rather velocity varies from zero at walls to maximum value at or near the centerline of pipe. Therefore, we define the average velocity V_{avg} as the average of v_n across the entire cross section.

Average velocity: $V_{avg} = \frac{1}{A} \int_A (v_n \times dA_c)$; where A – area of cross sectional normal to flow direction

For incompressible flow or even for compressible flow where ρ is uniform across area A:

That is $\dot{m} = \rho \times A \times V_{avg}$

CONSERVATION OF MASS PRINCIPLE:-

The conservation of mass principle for a control volume can be expressed as "the net mass transfer to or from a control volume during a time interval Δt is equal to the net change in the total mass within the control volume during Δt.

That is

(Total mass entering control volume during Δt) – (total mass leaving control volume during Δt)

$\equiv$ (net change of mass within control volume during Δt)

$M_{in} - M_{out} \equiv \Delta M_{cv}$ (kg); where $\Delta M_{cv} = M_{final} - M_{initial}$

These equations are known as mass balance and are applicable to any control volume undergoing any kind of process. Also, the total mass within the control volume at any instant in time 't' is determined by integration

Total mass within control volume; $M_{cv} = \int_{cv} \rho \times dv$

MASS BALANCE FOR STEADY FLOW PROCESSES:-

During a steady flow process, the total amount of mass contained within a control volume does not change with time. Therefore, according to conservation of mass principle, total amount of mass entering a constant volume is equal to total amount of mass leaving it. That is, conservation of mass principle for a general steady flow system with multiple inlets and outlets can be expressed as

Steady flow: $\quad \sum_{in} \dot{m} = \sum_{out} \dot{m} \quad$ (kg/s)

For steady, incompressible flow; $\quad \sum_{in} \dot{V} = \sum_{out} \dot{V} \quad$ (m^3/s)

Important thing to remember is that there exists no "conservation of volume" principle. Therefore, volume flow rates into and out of a steady flow device may be different. This is due to change in density at inlet and outlet. For steady flow of liquids, volumes flow rates as well as mass flow rates remain nearly constant since liquids are essentially incompressible substance.

Chapter – 6

CONCEPT OF FLOW WORK

Control volumes involve mass flow across their boundaries and some work is required to push the mass into or out of the control volume. This work is known as flow work or flow energy. If fluid pressure is P and cross sectional area of the fluid element is A, the applied force F required is given by F = P x A.

To push the entire fluid into the control volume, this force must act through a distance L. Thus, the workdone in pushing the fluid element across the boundary

W_{flow} = F x L = P x A x L = P x V   (kJ)

ENERGY ASSOCIATED WITH FLOWING FLUID: -

The total energy of a simple compressible system consists of three parts: internal, kinetic and potential energies. Therefore, on a unit mass basis,

$e = u + k.e. + p.e. = u + \frac{v^2}{2} + g \times z$ (kJ/kg), This is valid for non flowing fluid.

The fluid entering or leaving a control volume possesses an additional form of energy – flow energy (P x v)

$\Phi = P \times v + e = P \times v + (u + k.e. + p.e.) = (u + Pv) + k.e. + p.e. = h + \frac{v^2}{2} + g \times z$ (kJ/kg)

ENERGY TRANSFER ASSOCIATED WITH MASS TRANSFER: -

Note that Φ is total energy per unit mass, therefore the total energy of a flowing fluid of mass m is simply mΦ, assuming the property of mass 'm' are uniform.

Amount of energy transfer by mass: - $E_{mass} = m\Phi = m (h + \frac{v^2}{2} + g \times z)$ (kJ)

ENERGY ANALYSIS OF STEADY FLOW SYSTEM: -

Steady flow process may be defined as a process during which a fluid flows through a control volume steadily. That is, fluid property can change from point to point within the control volume but at any point, they remain constant during the entire process. During steady flow process, no intensive or extensive property within the control volume change with time. Thus, volume V, mass m and total energy content E of control volume remains constant. Therefore, boundary work is zero for steady flow process.

Also, the heat and work interaction between steady flow system and its surrounding do not change with time. Thus, the power delivered by a system and rate of heat transfer to or from a system remains constant during steady flow process. Thus, total mass or energy entering the control volume must be equal to the total mass or energy leaving it.

The mass balance for steady flow system: $\sum_{in} \dot{m} = \sum_{out} \dot{m}$ (kg/s)

Energy balance for steady flow system: $\dot{E}_{in} - \dot{E}_{out} = \frac{dE_{system}}{dt} = 0$

$\Rightarrow \quad \dot{E}_{in} = \dot{E}_{out}$ (kW)

IMPORTANT STEADY FLOW ENGINEERING DEVICE: -

Nozzle and diffuser: - A nozzle is a device that increases the velocity of a fluid at the expense of pressure and diffuser is a device that increases pressure of a fluid by slowing it down. The cross sectional area of nozzle decreases in the flow direction for subsonic flows and for supersonic flows and the reverse is true for diffuser. The rate of heat transfer between the fluid flowing through a nozzle or diffuser and the surrounding is usually very small since fluid has high velocities and thus, it does not spend enough time in the device for any significant heat transfer to take place.

Nozzle and diffuser typically involve no work and any change in potential energy is negligible. But nozzle and diffuser usually involve very high velocities and a fluid passes through a nozzle or diffuser, it experiences large changes in its velocity. Therefore, the kinetic energy changes must be account for analyzing the flow.

Turbine and compressor: - As the fluid passes through the turbine, work is done against the blades, which are attached to the shaft. As a result, the shaft rotates and turbine produces work. Compressor, pump and fans are devices used to increase the pressure of a fluid. Work is supplied to these devices from an external source through a rotating shaft. Therefore, compressor involves work input. Even though these three devices function similarly but they do differ in the task they perform. A fan increases the pressure of a gas slightly and is mainly used to mobilize a gas. A compressor is capable of compressing the gas to very high pressure. Pumps work very much like compressor except that they handle liquids instead of gases.

Heat transfer from turbine is usually negligible since they are typically well insulated. Heat transfer is also negligible for compressor. Potential energy changes are negligible for all these devices. The velocity involved in these devices with exception of turbines and fans are usually too low to cause any significant change in kinetic energy.

The fluid velocities encountered in most turbine are very high and fluid experience a significant change in its kinetic energy but this change is usually very small relative to the change in enthalpy and thus it is often disregarded.

Throttling valves: - Throttling devices are any kind of devices that cause a significant pressure drop in the fluid. Unlike turbines, they produce pressure drop without involving any work. Pressure drop in fluid causes large drop in temperature and because of this they are used in refrigeration and air conditioning application. The magnitude of temperature rise or drop during a throttling process is governed by property called Joule Thomson coefficient.

Chapter – 7

SECOND LAW OF THERMODYNAMICS

Important terminologies: -

Thermal energy reservoirs: - A hypothetical body with relatively large thermal energy capacity that can supply or absorb finite amount of heat without undergoing any change in temperature. Such a body is called thermal energy reservoir.

A reservoir that supplies energy in the form of heat is called a **source** and reservoir that absorb energy in the form of heat is called **sink.**

Heat engines: -

Heat engines are those devices which convert heat energy into work energy. They receive heat from a high temperature source and convert part of it into work and rest is rejected to low temperature sink. They operate on a cycle. Heat engines and other cyclic devices usually involve a fluid to and from which heat is transferred while undergoing a cycle. This fluid is called working fluid. Common example of heat engine is steam power plant.

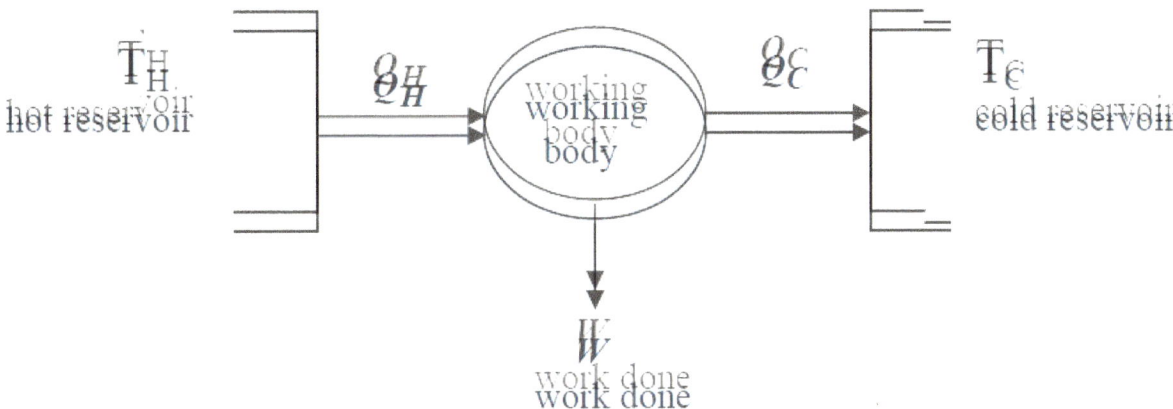

Let,

Q_{in} ≡ amount of heat supplied from high temperature source

Q_{out} ≡ amount of heat rejected to low temperature sink

W_{out} ≡ amount of work delivered or output

W_{in} ≡ amount of required input

Net workdone obtained = total work output − total work input

$$W_{net,out} = W_{out} - W_{in}$$

While for a closed system undergoing a cycle, the change in internal energy ΔU is zero and therefore, net work output of system is also equal to net heat transfer to the system

$$W_{net,out} = Q_{in} - Q_{out} \qquad (kJ)$$

THERMAL EFFICIENCY: -

It is the ratio of work output to heat input. In other words, the fraction of the heat input heat that is converted to net work output is a measure of the performance of a heat engine and is called thermal efficiency η_{th}.

$$\text{Thermal efficiency} = \frac{\text{net work output}}{\text{total heat input}}$$

$$\Rightarrow \eta_{th} = \frac{W_{net,out}}{Q_{in}} = 1 - \frac{Q_{out}}{Q_{in}}$$

The thermal efficiency of a heat engine is always less than unity since both Q_{in} and Q_{out} are defined as positive quantities.

KELVIN PLANCK'S STATEMENT: SECOND LAW OF THERMODYNAMICS

It states that "it is impossible for any device that operates on a cycle to receive heat from a single reservoir and produce a net amount of work". That is heat engine must exchange heat with a low temperature sink as well as high temperature source to keep operating.

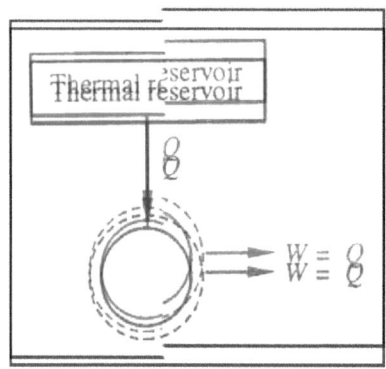

In other words, "no heat engine can have a thermal efficiency of 100 percent".

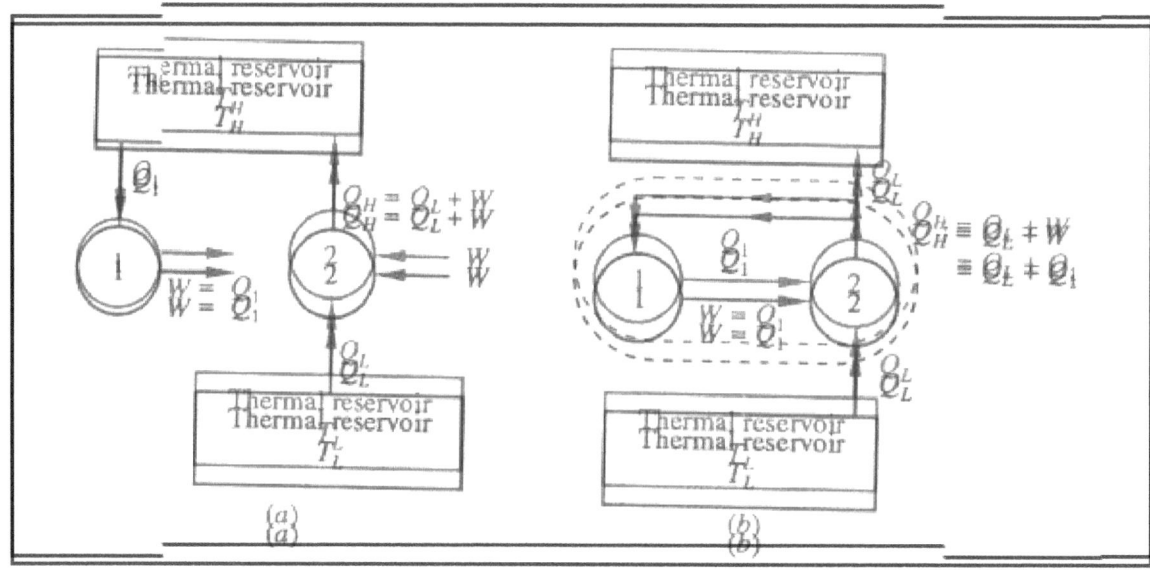

CLAUSIUS STATEMENT: SECOND LAW OF THERMODYNAMICS

It states that: "it is impossible to construct a device that operates in a cycle and produces no effect other than the transfer of heat from a lower temperature body to a higher temperature body".

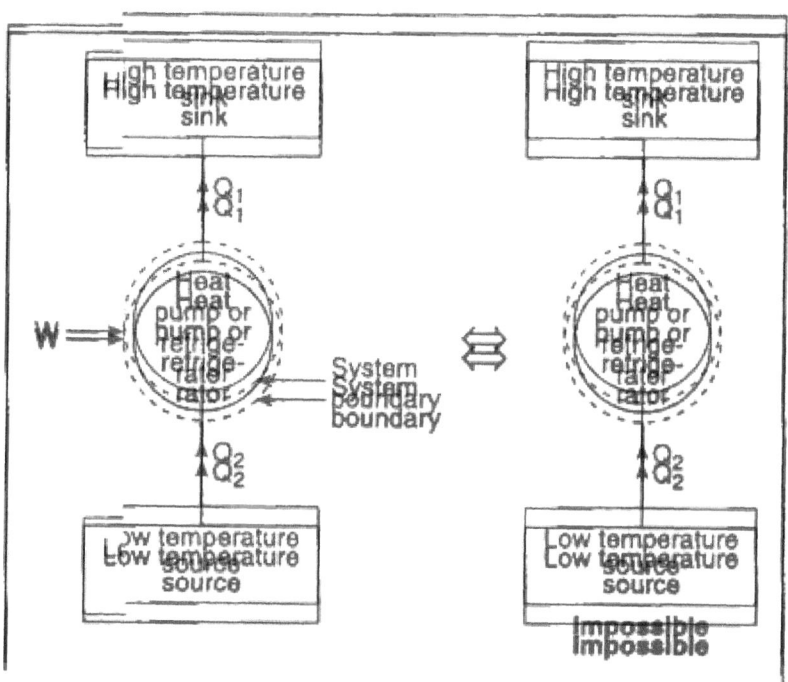

Both Kelvin-Planck and Clausius statement of second law are negative statement and a negative statement cannot be proved.

REFRIGERATORS AND HEAT PUMPS: -

The transfer of heat from a low temperature medium to high temperature medium requires a device is called refrigerators and heat pump. Refrigerators and heat pumps are cyclic devices. The working fluid used in the refrigeration cycle is called refrigerant.

In case of refrigerators, refrigerants enter compressor as vapor and is compressed to the condenser pressure. It leaves the compressor at relatively high temperature and cools down and condenses as it flows through the coil of condenser by rejecting heat to the surrounding medium. It then enters a capillary tube where its pressure and temperature drop drastically due to throttling effect.

The low temperature refrigerant then enters the evaporator, where it evaporates by absorbing heat from the refrigerated space. The cycle is completed as the refrigerants leaves the evaporator and re - enters the compressor.

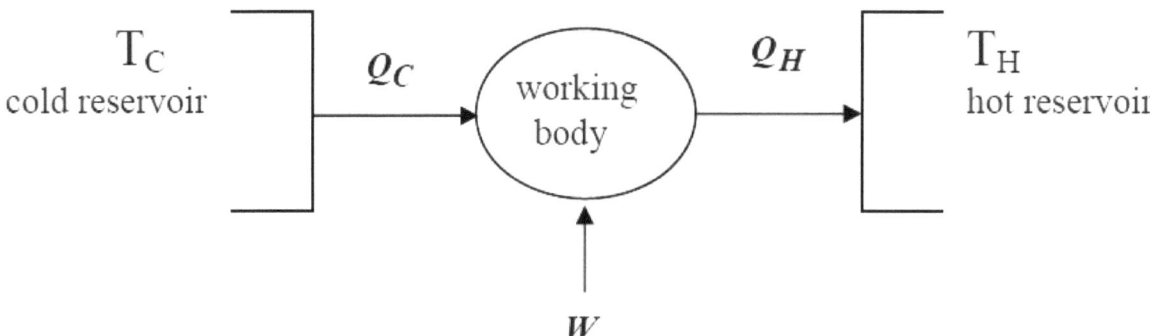

Although, it looks that refrigerator and heat pump are same, but it is not so. They differ in their function like refrigerators objective is to maintain a confined space at low temperature while heat pump objective is to maintain a heated space at high temperature. This is accomplished by absorbing heat from a low temperature source and supplying this heat to high temperature medium.

COEFFICIENT OF PERFORMANCE: -

The efficiency of a refrigerator and heat pump are expressed in terms of coefficient of performance (COP).

COP of refrigerator is given by, $COP_R = \dfrac{\text{desired output}}{\text{required input}} = \dfrac{Q_L}{W_{net,in}} = Q_L / (Q_H - Q_L)$

The value of COP_R can be greater than unity. That is, the amount of heat removed from the refrigerated space can be greater than the amount of work input.

COP of heat pump is given by, COP_{HP} = desired output / required input = $Q_H / W_{net,\,in}$

Also, $COP_{HP} = Q_H / (Q_H - Q_L) = 1 / \left[1 - \left(\dfrac{Q_L}{Q_H}\right)\right]$

⇨ $COP_{HP} = COP_R + 1$ (for fixed value of Q_L and Q_H)

This relation shows that COP_{HP} is always greater than unity since COP_R is a positive quantity. That is, heat pump will function, at worst, as a resistance heater, supplying as much energy to the confined space as it consumes.

PERPETUAL MOTION MACHINE: -

Any device that violates either law is called a perpetual motion machine. A device that violates the first law of thermodynamics (by creating energy) is called perpetual motion machine of the first kind (PMM1) and a device that violates the second law of thermodynamics is called perpetual motion machine of second kind (PMM2).

REVERSIBLE AND IRREVERSIBLE PROCESSES: -

A reversible process is defined as a process that can be reversed without leaving any trace on the surrounding. That is, both the system and surrounding are returned to their initial states at the end of the reverse process. This is possible only if the net heat and net work exchange between the system and the surrounding is zero for the combined process (original and reverse process).

Processes that are not reversible are called irreversible processes. Generally we assume a process as reversible because they are easy to analyze, since a system passes through a series of equilibrium states during a reversible process. They can be served as ideal model to which actual processes compared.

Irreversibility: - the factors that cause a process to be reversible are called irreversibilities. They include friction, unrestrained expansion, mixing of two fluids, heat transfer across a finite temperature difference, inelastic deformation of solids and chemical reactions.

INTERNALLY AND EXTERNALLY REVERSIBLE PROCESSES: -

A process is called internally reversible if no irreversibilities occur within the boundaries of the system during the process. During an internally reversible process, a system proceeds through a series of equilibrium states and when the process is reversed, the system passes through exactly the same equilibrium states while returning to its initial state. The quasi equilibrium process is an example of an internally reversible process.

A process is called externally reversible if no irreversibilities occur outside the system boundaries during the process. Heat transfer between a reservoir and system is an externally reversible process if the outer surface of system is at the temperature of the reservoir.

A process is called totally reversible or simply reversible, if it involves no irreversibilities within the system or its surrounding. A totally reversible process involves no heat transfer through a finite temperature difference, no quasi equilibrium changes and no irreversibilities associated.

Chapter – 8

CARNOT CYCLE

The Carnot cycle is composed of two reversible processes – two isothermal and two adiabatic processes. The theoretical heat engine that operates on the Carnot cycle is called Carnot heat engine.

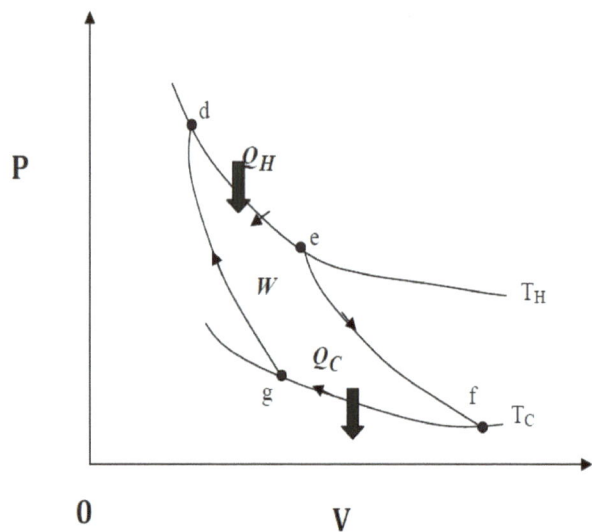

Main processes are as under:-

de = Reversible isothermal expansion process

ef = Reversible adiabatic expansion

fg = Reversible isothermal compression process

gd = Reversible adiabatic compression process

REVERSED CARNOT CYCLE: -

All the processes that comprise Carnot heat engine cycle can be reversed and in this case it becomes Carnot reversed cycle or Carnot refrigeration cycle. The cycle remains exactly the same, except that the direction of any heat and work interaction is reversed.

CARNOT PRINCIPLES: -

Thermal efficiency of reversible and irreversible heat engine are based on two Carnot principles known as Carnot principle.

The efficiency of an irreversible heat engine is always less than the efficiency of a reversible heat engine operating between same two reservoirs.

The efficiencies of all reversible heat engines operating between same two reservoirs are the same.

THERMODYNAMIC TEMPERATURE SCALE: -

A temperature scale that is independent of the property of the substance that are used to measure temperature is called a thermodynamic temperature scale.

The second Carnot principle states that all reversible heat engines have the same thermal efficiency when operating between the same two reservoirs. That is, the efficiency of a reversible engine is independent of the working fluid and its properties, the way the cycle is executed or type of reversible engine used.

Since the energy reservoirs are characterized by their temperature, the thermal efficiency of reversible heat engine is a function of the reservoir temperature only. That is,

$$\eta_{th, rev} = f(T_H, T_L) \quad \text{or} \quad Q_H/Q_L \equiv g(T_H, T_L)$$

$$\Rightarrow (Q_H/Q_L)_{rev} \equiv T_H/T_L$$

The general thermodynamic which we used today is known as Kelvin temperature scale and temperature on this scale is called absolute temperature. On the Kelvin temperature scale, temperature ratios depend on the ratios of heat transfer between a reversible heat engines and reservoirs and are independent of the physical property of any substance. On this scale, temperatures vary between zero and infinity.

Also, $T(^\circ C) = T(^\circ K) - 273.15$

IMPORTANT POINTS TO REMEMBER:

Carnot Heat engine-

The thermal efficiency of any heat engine, reversible or irreversible is given as

$$\eta_{th} = 1 - (Q_L/Q_H)$$

This relation is also known as Carnot efficiency. This is the highest efficiency a heat engine operating between the two thermal energy reservoirs at temperature T_L & T_H can have.

Also,

$\eta_{th} < \eta_{th,rev}$ irreversible heat engine
$\eta_{th} = \eta_{th,rev}$ reversible heat engine
$\eta_{th} > \eta_{th,rev}$ impossible heat engine

The η_{th} of actual heat engine can be maximized by supplying heat to the engine at the highest possible temperature and rejected heat from the engine at the lowest possible temperature.

Chapter – 9

ENTROPY

The important inequality in thermodynamics that has major consequences is the Clausius inequality $\oint \delta Q/T \leq 0$. That is, the cyclic integral of $\delta Q/T$ is always less than or equal to zero. This inequality is valid for all cycles, reversible or irreversible.

Also, the equality in the Clausius inequality holds for totally or just internally reversible cycles and the inequality for the irreversible cycles.

We know that a quantity whose cyclic integral is zero depends on the state only and not the process path and thus it is a property. Therefore, the quantity $(\delta Q/T)_{int\ rev}$ must represent a property. Thus, the name of this property is entropy and designated by "S". Entropy is an extensive property of a system.

$dS = (\delta Q/T)_{int\ rev}$ (kJ/K)

ENTROPY GENERATION: -

Consider a cycle that is made up of two process, process 1 – 2, which is arbitrary (reversible or irreversible) and process 2 – 1, which is internally reversible.

From Clausius inequality, $\oint \delta Q/T \leq 0$

$\Rightarrow \int_1^2 (\delta Q/T) + \int_2^1 (\delta Q/T)_{int\ rev} \leq 0$

$\Rightarrow \int_1^2 (\delta Q/T) + S2 - S1 \leq 0$

$\Rightarrow S_2 - S_1 \geq \int_1^2 (\delta Q/T) \quad \text{or} \quad dS \geq (\delta Q/T)$

Where equality holds for an internally reversible process and inequality for an irreversible process. Also, entropy change of a closed system during an irreversible process is greater than the integral of $(\delta Q/T)$ evaluated for that process.

Also, some entropy is generated or created during an irreversible process and this generation is due to presence of irreversibilities. The entropy generated during a process is called entropy generation and denoted by S_g. Note that the difference between entropy change of a closed system and entropy transfer is equal to entropy generation.

That is, $\Delta S_{sys} = S_2 - S_1 = \int_1^2 (\delta Q/T) + S_g$

Also, S_g is always positive quantity or zero. Its value depends on the process and thus it is not a property of system.

INCREASE OF ENTROPY PRINCIPLE: -

For an isolated system (or adiabatic closed system), heat transfer is zero,

⇨ $\Delta S_{isolated} \geq 0$

This equation can be expressed as the entropy of an isolated system during a process always increases or in limiting case of a reversible process remains constant. In other words, it never decreases. This is known as increase of entropy principle.

Note that in the absence of any heat transfer, entropy change is due to irreversibilities only and their effect is always to increase entropy.

Since no actual process is truly reversible, we conclude that some entropy is generated during process and therefore, entropy of universe, which can be considered to be an isolated system, is continuously increasing.

Important points to remember: -

Processes can occur in a certain direction only. A process must proceed in the direction that complies with increase of entropy principle. That is, $S_g \geq 0$. A process that violates this principle is impossible.

The performance of engineering system is degraded by the presence of irreversibilities and entropy generation is a measure of the magnitude of irreversibilities present during that process.

Entropy is conserved during the idealized reversible processes only and increase during all actual processes.

The entropy of a fixed mass can be changed by heat transfer and irreversibilities. A process during which the entropy remains constant is called an isentropic process. Also, a reversible adiabatic process is definitely isentropic but an isentropic process is not necessarily a reversible adiabatic process.

THIRD LAW OF THERMODYNAMICS: -

Entropy can be viewed as a measure of molecular disorder or molecular randomness. The entropy of a system increases whenever the molecular randomness or uncertainty of a system increases.

"The entropy of a pure crystalline substance at absolute zero temperature is zero since there is no uncertainty about state of molecules at that instant". This statement is known as third law of thermodynamics.

It provides an absolute reference point for the determination of entropy. The entropy determined relative to this point is called absolute entropy.

GIBBS EQUATION

The differential form of the conservation of energy equation for a closed stationary system containing a simple compressible substance can be expressed for an internally reversible process as

$\delta Q_{int\ rev} - \delta W_{int\ rev,\ out} = dU$

But $\delta Q_{int\ rev} = TdS$ and $\delta W_{int\ rev,\ out} = Pdv$

Thus, $TdS = dU + Pdv$ (kJ)

This is known as first TdS or Gibbs equation.

Note that the only type of work interaction a simple compressible system involve as it undergoes an internally reversible process is the boundary work.

The second TdS equation is obtained by eliminating dU from above equation by using the definition of enthalpy ($H = U + Pv$)

$H = U + Pv$

$\Rightarrow dH = dU + Pdv + vdP$

we know that $TdS = dU + Pdv$

$\Rightarrow dH = TdS + vdP$

$\Rightarrow TdS = dH - vdP$

Both equations are property relation and therefore are independent of the type of the processes. However, equation obtained are valid for both reversible and irreversible processes since entropy is a property and change in a property between two states is independent of the type of process the system undergoes.

Entropy change of incompressible substance: -

Liquid and solids can be approximately as incompressible substances since their specific volumes remain nearly constant during a process. Thus, $dv = 0$ for solids and liquids,

$dS = dU/T = CdT/T$ (since $C_p = C_v = C$ and $dU = CdT$ for incompressible substance)

Then, entropy change during a process is determined by integration

$$S_2 - S_1 = \int_1^2 C(T) \frac{dT}{T} = C_{avg} \ln\frac{T_2}{T_1} \quad (kJ)$$

Where C_{avg} = average specific heat of the substance over the given temperature interval.

Note that the entropy change of truly incompressible substance depends on temperature only and is independent of pressure. The effects of volume should be considered when temperature change is large.

For isentropic process, $S_2 - S_1 = C_{avg} \ln\frac{T_2}{T_1} = 0$

$\Rightarrow T_1 = T_2$

That is temperature of truly incompressible substance remains constant during an isentropic process. Therefore, the isentropic process of an incompressible substance is also isothermal.

Entropy change of ideal gases: -

Case – 1: we know $dS = \dfrac{dU}{T} + \dfrac{Pdv}{T}$

Also, $dU = C_v dT$ and $P = \dfrac{RT}{V}$

$$\Rightarrow dS = C_v \dfrac{dT}{T} + R \dfrac{dv}{V}$$

Therefore, the entropy change for a process is obtained by integrating

$$S_2 - S_1 = \int_1^2 C_v(T) \dfrac{dT}{T} + R \ln \dfrac{V_2}{V_1}$$

Case – 2: similarly, $dS = \dfrac{dH}{T} - v\dfrac{dP}{T}$

And $dH = C_p dT$; $V = \dfrac{RT}{P}$

$$\Rightarrow dS = C_p \dfrac{dT}{T} - R \dfrac{dP}{P}$$

On integrating, $S_2 - S_1 = \int_1^2 C_p(T) \dfrac{dT}{T} - R \ln \dfrac{P_2}{P_1}$

Isentropic processes of ideal gases

We know, $dS = C_v \ln\frac{T_2}{T_1} + R \ln\frac{v_2}{v_1} = 0$

$\Rightarrow \ln\frac{T_2}{T_1} = -\frac{R}{C_v} \ln\frac{v_2}{v_1}$

Assuming constant specific heat,

Then, $\ln\frac{T_2}{T_1} = \ln\left(\frac{v_2}{v_1}\right)^{R/C_v}$

$\Rightarrow \frac{T_2}{T_1} = \left(\frac{v_2}{v_1}\right)^{k-1}$; where k = specific heat ratio

Similarly, $\frac{T_2}{T_1} = \left(\frac{P_2}{P_1}\right)^{(k-1)/k}$

From above two relations, we get, $\frac{P_2}{P_1} = \left(\frac{v_2}{v_1}\right)^k$

These above equation are valid for isentropic process only when constant specific heat assumption is considered.

REVERSIBLE STEADY FLOW WORK: -

We know that $W = -\int_1^2 vdP$

That is the reversible steady flow work is closely associated with the specific volume of fluid flowing through the device. The larger the specific volume the larger the reversible work produced or consumed by steady flow device.

In steam power plant, pump handle liquid, which has a very small specific volume and turbine handle vapor, whose specific volume is many times larger. Therefore, the work output of turbine is much larger than the work input to pump.

If we have to compress the steam exiting the turbine back to the turbine inlet pressure before cooling it first in the condenser in order to save the heat rejected, we would have to supply all the work produced by turbine back to compressor. In reality, required work input would be even greater than the work output of the turbine because of irreversibilities present in both processes.

Also, to reduce the compressor work, we need to keep the specific volume of the gas as small as possible during compression process. This is done by maintaining the temperature of gas as low as possible during compression since the specific volume of gas is proportional to temperature. Therefore, reducing the work input to a compressor requires that the gas be cooled as it is compressed.

REFERENCES: -

Adams, Steve & Halliday, Jonathan, (2000). *Advanced Physics*. UK: Oxford University Press.

Francis, W. Sears & Gerhard, L. Salinger. (1975). *Thermodynamics, Kinetic Theory, and Statistical Thermodynamics (Third Edition)*, Philippines: Addison-Wesley Publishing Company Inc..

Halliday, David., Resnick, Robert & Walker, Jeanrl. (2001). *Fundamentals of Physics-Extended (Sixth edition)*. New York: John Wiley and Sons, Inc.

John, D. Cutnell & Kenneth, W. Johnson (1989). *Physics*. USA: John Wiley & Sons, Inc.

Mark, W. Zemansky & Richard, H. Dittmann (1981*).Heat and Thermodynamics* (Sixth Edition). New York: McGraw-Hill Book Company.

Stuart, B. Palmer & Mircea, S. Rogalski (1996). *Advance University Physics*. SA: Gordon and Breach Science Publisher.

Rao, Y.V.C. (2004). *An Introduction to Thermodynamics* (Revised Edition).India: University Press (India) Private Limited.

Cengel & boles, Thermodynamics : an engineering approach, TMH

www.ingramcontent.com/pod-product-compliance
Lightning Source LLC
Chambersburg PA
CBHW051917210526
45473CB00006B/2044